GERMAN HEAVY FIELD ARTILLERY

Reloading an sFH 18 on the Eastern Front (summer 1943) following a short bombardment – hard work in the hot sun!

1934 - 1945

Joachim Engelmann

Schiffer Military/Aviation History
Atglen, PA

Changing positions in the Westwall (Siegfried Line to the Allies) on April 15, 1940. Here the howitzer's barrel has just been loaded from the gun carriage trailer (left) on to the gun barrel trailer (right). Members of the gun crew are pushing the gun barrel trailer.

Translated from the German by David Johnston

Printed in the United States of America.
ISBN: 0-88740-759-5

This title was originally published under the title,
Waffen-Arsenal Waffen und Fahrzeuge der Heere und Luftstreitkräfte Deutsche Schwere Feldhaubitzen 1934-1945,
by Podzun-Pallas-Verlag, Friedberg.

Published by Schiffer Publishing Ltd.
77 Lower Valley Road
Atglen, PA 19310
Please write for a free catalog.
This book may be purchased from the publisher.
Please include $2.95 postage.
Try your bookstore first.

We are interested in hearing from authors with book ideas on related topics.

From the Foot Artillery to the "Heavy Artillery of the Field Army"

After the Franco-Prussian war of 1870-71, in which the German artillery had proved its worth, on May 7, 1874 Inspector-General of Artillery Generalleutnant von Podbielski separated the foot artillery from the field artillery and organized it as a second, independent branch of the artillery. However, on account of its immobility, until 1891 it remained primarily fortress artillery with no role in the field, even though it was equipped with 120-150mm cannon and 150-210mm howitzers. The foot artillery so eagerly demanded by Kaiser Wilhelm II and Schlieffen's Chief Quartermaster von Deines, namely one battalion equipped with 150mm howitzers and 210mm large-bore howitzers per army corps, took part in its first field exercise on August 12-13, 1887. This was the work of the first Inspector-General of Foot Artillery, General der Artillerie von Roerdansz. In autumn 1888 there followed a mock attack on Fort Rheinhell in Koblenz from the Lahn Valley. There were live firing exercises by individual batteries at Küstrin in 1889, at Lötzen and Nideggen in 1890 and at Haltern in 1891. These firing camps subsequently became regular annual affairs. The six-meter-deep mining effect of the recently-developed 210mm Schiesswollgranate (Guncotton Shell) C/83, 1050mm long, had ended the mission of fortress warfare for good. There was still a role for mobile heavy guns, but not for stationary fortress weapons.

The difficulties of the far-reaching changes and new development lay in the following: in the past draught horses had been rented whenever the guns had to be moved; now the units had to keep and maintain their own stable of horses, for now they were concerned not just with transport, but with mobility in open country. The firing height in the gun carriage was lowered significantly to achieve improved durability. The gun bases, for so long considered indispensable, which prolonged positioning and readying of the guns to up to 1 1/2 hours, were finally abandoned after comparative firing trials in 1896. Heinrich Ehrhardt's revolutionary invention of barrel recoil with the hydraulic barrel recoil brake of 1899 pointed the way to the modern artillery weapon. Even in the foot artillery, riding, driving and firing were replaced by universal scientific artillery training. The next significant design to incorporate modern technical developments was the sFH 18, developed in the period 1926 to 1930 and introduced into service in 1933-34. A fundamental change had taken place in the twenty productive years before World War One: the foot artillery had become mobile. Weapons training was now carried out at the artillery's own practice ranges at Thorn and Wahn from concealed firing positions; concentrations of fire by battalions and regiments within twenty minutes became commonplace. The first war experience with a heavy field howitzer battery was gained in the campaign in China in 1900. Five new heavy field howitzer battalions were activated in 1902-1903, following a reduction in company strengths from 250 to 200 men. Not until November 19, 1908 were the companies given the right to use the designation "battery." Prior to the outbreak of war marksmanship and tactical training at the division level had attained a high standard of performance. Observers, flash-ranging, tethered balloons, long-range photography, photogrammetry and, from 1913, observers in aircraft were used for artillery suppression. The great fragmentation effect of the Granate 04 made it an extremely effective shell. A request for three air observation battalions for the artillery was turned down. Three more artillery regiments were activated in 1912 and two more were planned. In July 1914, just before the start of the war, there were 25 heavy regiments each with two battalions and one or two horse-drawn battalions, of which at first only 32 were field mobile, as in peacetime the batteries were not incorporated in the army corps. These units had a total of 840 sFH 02 and 870 sFH heavy howitzers, which, beginning in October 1914, were replaced by the sFH 13. This represented a considerable offensive superiority over the enemy. Mobilization, which took place on August 2, resulted in an increase of 75 heavy field howitzer battalions, divided almost equally between active, reserve and regional defense units. The final tally was 426 battalions in 63 regiments. When the war ended all heavy field howitzers had to be surrendered or destroyed, except for twelve sFH 13 lg. with 400 rounds per gun at Königsberg. The total amount of heavy artillery surrendered to the victorious powers was 1,566 batteries with 4,826 guns!

The sFH 13 reached the units just in time, a few weeks before the outbreak of the First World War. As the most modern artillery piece of the day, it immediately played an important role in the German Army's operations. Its advantages compared to its predecessors were obvious. The new combination of barrel recoil brake and compressed air, which improved the weapon's stability, attracted special attention and demanded special servicing procedures. These were incorporated into service manuals which were printed until 1918 and which contained detailed technical information and thorough instructions for handling the weapon. The technology was even more sensitive than it is today.

sFH 13 Specification

Country of manufacture		*Germany*
Manufacturer		*Krupp*
Designation		*Heavy Field Howitzer 13*
Caliber	*mm*	*149.7*
Barrel length	*calibers*	*14*
Sighting range, elevation	*degrees*	*– 0 + 45*
Sighting range, azimuth	*degrees*	*5*
Weight in firing position	*kg*	*2135*
Projectile weight	*kg*	*42*
Muzzle velocity	*m/sec*	*365*
Maximum range	*km*	*8.5*
Gun carriage		*box-trail*
Protective shield		*yes*
Means of locomotion		*horse-drawn*
In service in		*German Empire, Turkey*
Notes		*Transported in one piece. Also built by Rheinmetall.*

Rücklaufbremse der s. F. H. 13.

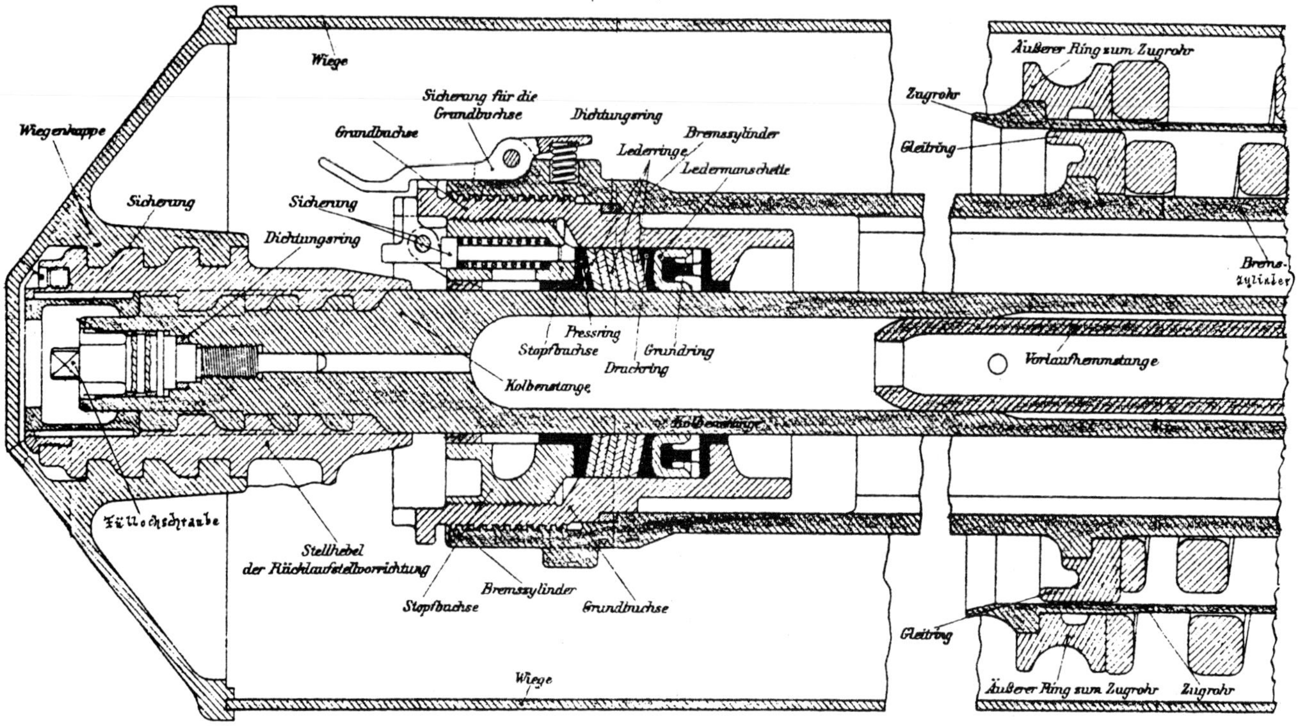

Cross-section of the recoil brake from ahead and behind.

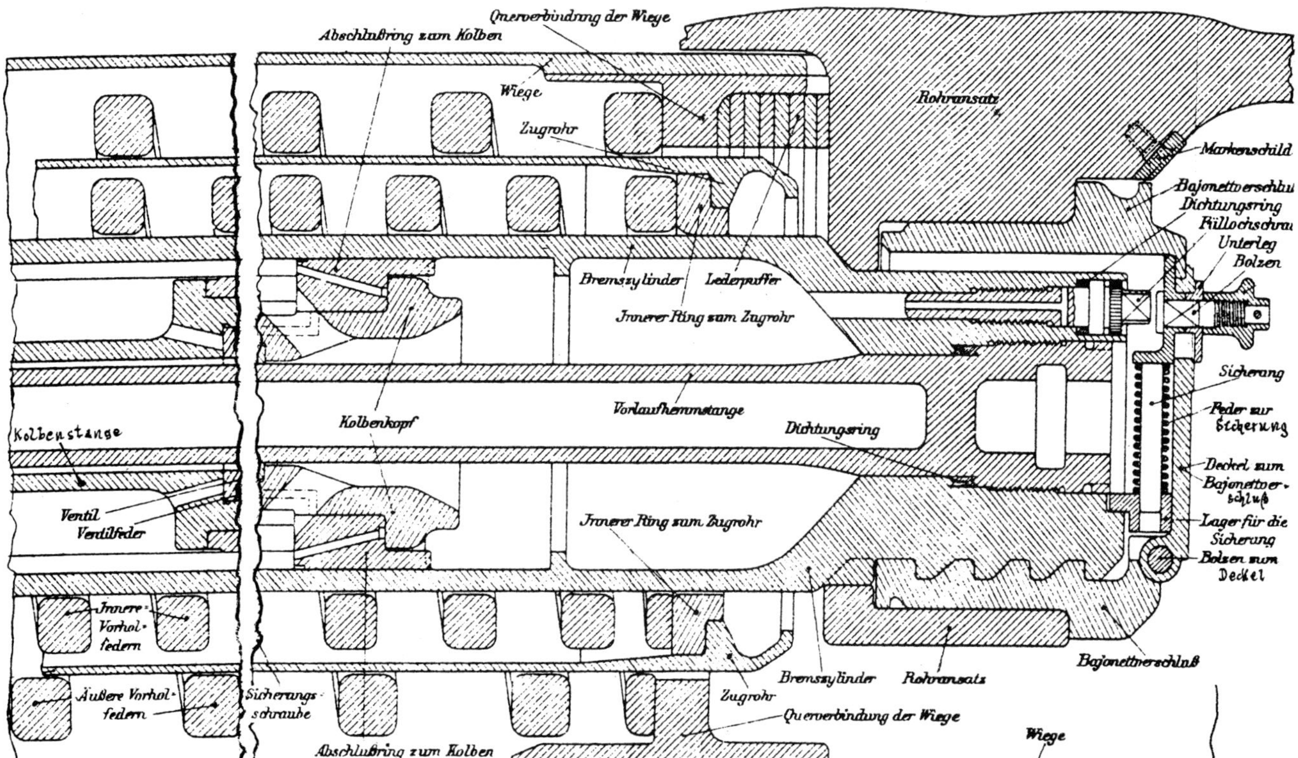

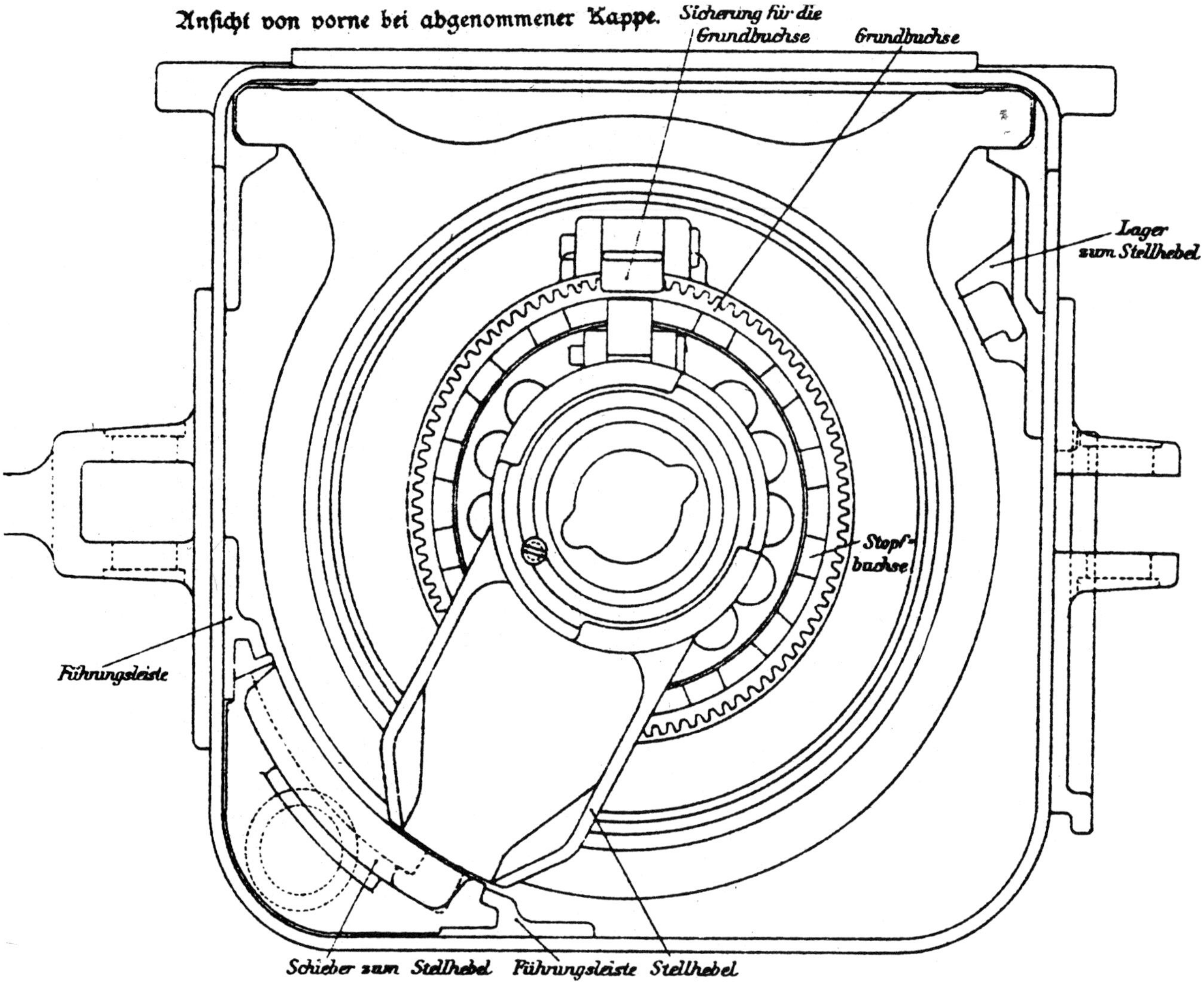

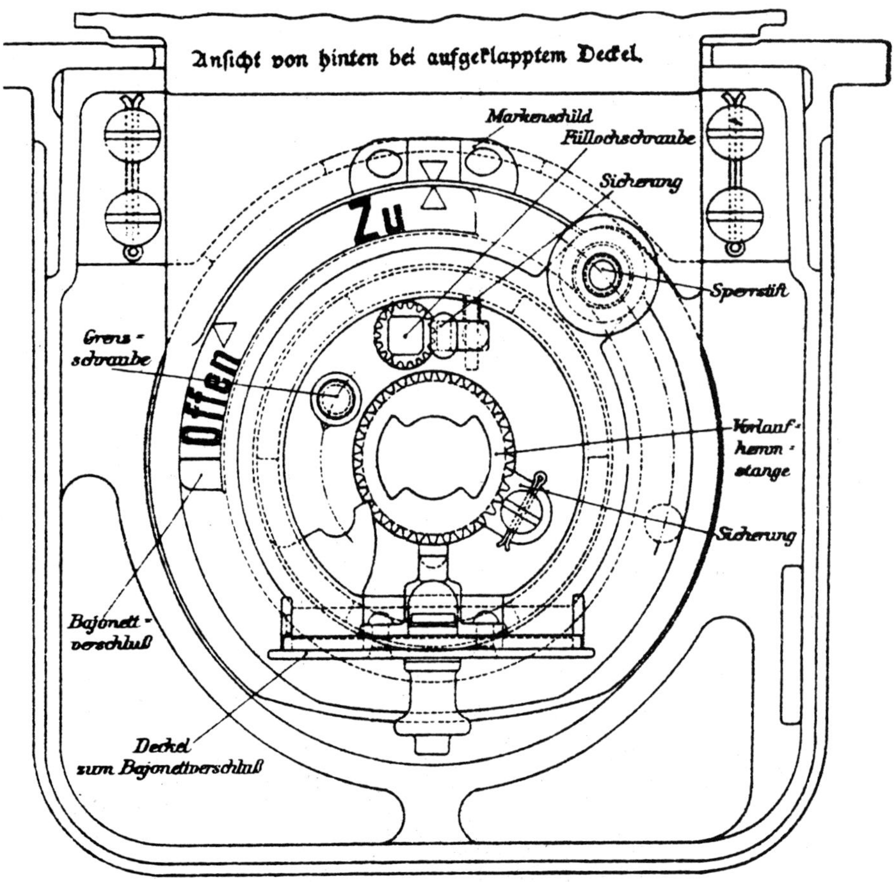

Cross-section of the compressed air recuperator from in front and behind.

Fully-limbered sFH 13, now "mobile artillery" instead of mere "foot artillery" as in the beginning.

The compressed air recuperator of the sFH 13.

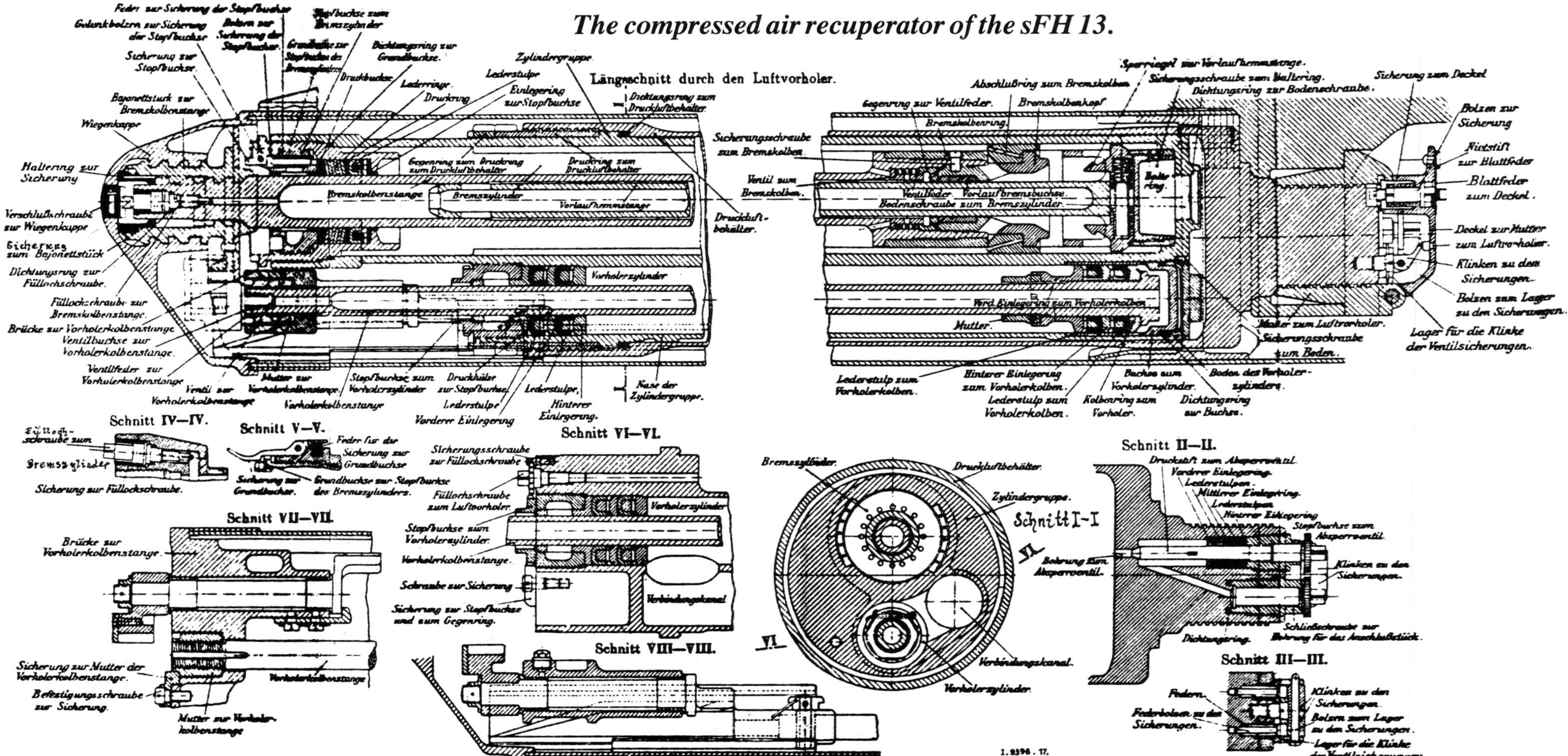

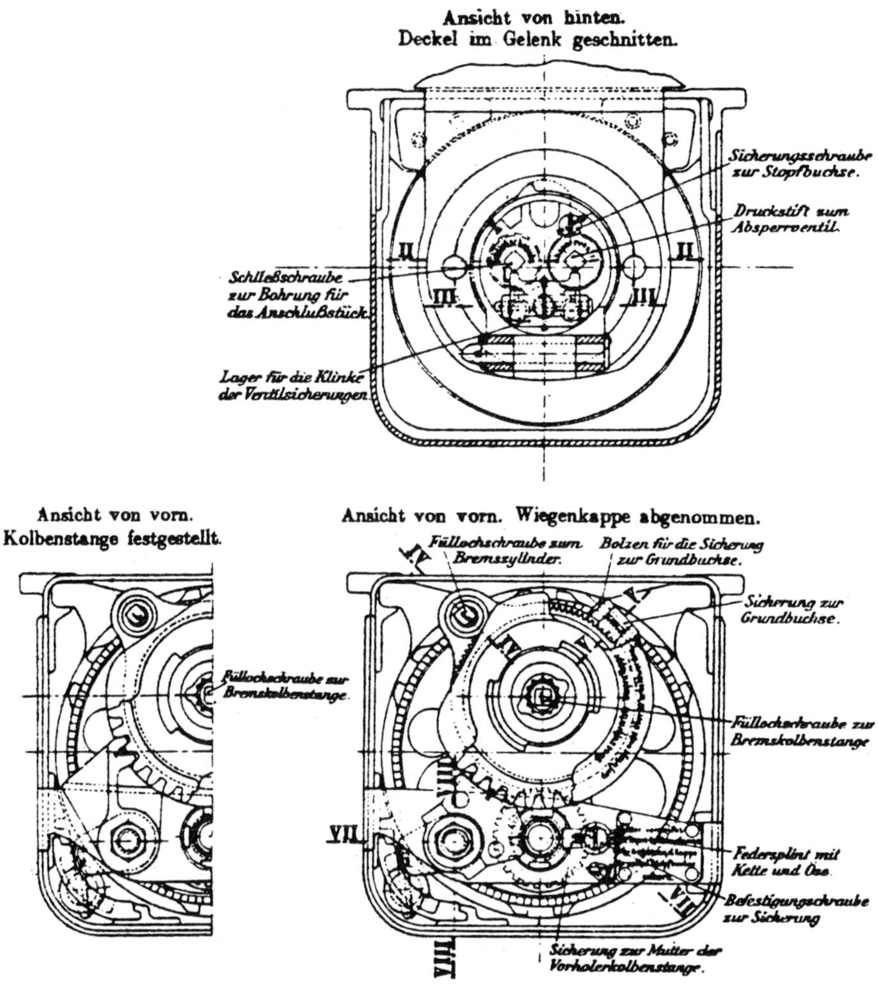

Country of manufacture		*German Empire*
Manufacturer		*Krupp*
Designation		*150mm-long Heavy Field Howitzer 13/02*
Caliber	*mm*	*149.7*
Barrel length	*calibers*	*17*
Sighting range, elevation	*degrees*	*– 0 + 45*
Sighting range, azimuth	*degrees 5*	
Weight in firing position	*kg*	*2210*
Projectile weight	*kg*	*42/44*
Muzzle velocity	*m/sec*	*365/377*
Maximum range	*km*	*8.5/8.8*
Gun carriage		*box-trail*
Protective shield		*yes*
Locomotion		*horse-drawn*
In service in		*German Empire*
Notes		*Transported in one piece.*

Long-barrelled Heavy Field Howitzer 13 L/13 seen from the left side. The lengthened barrel is quite obvious.

Long sFH 13

Long-barreled Heavy Field Howitzer 13. Seen from the right front, it is the picture of a powerful, compact piece.

Country of manufacture		German Empire
Manufacturer		Krupp
Designation		150mm-long Heavy Field Howitzer 13
Caliber	mm	149.7
Barrel length	calibers	17
Sighting range, elevation	degrees	– 0 + 45
Sighting range, azimuth	degrees	5
Weight in firing position	kg	2250
Projectile weight	kg	42/44
Muzzle velocity	m/sec	365/377
Maximum range	km	8.5/8.8
Gun carriage		box-trail
Protective shield		yes
Locomotion		drawn by horse or motor vehicle
In service in		German Empire, Begium, Latvia, Lithuania, the Netherlands
Notes		Transported in one piece. Also built by Rheinmetall.

The performance data of the "Long sFH 13" shows a well-thought-out design.

Country of manufacture		German Empire
Manufacturer		Krupp
Designation		150mm Heavy Field Howitzer M 17
Caliber	mm	149.7
Barrel length	calibers	14
Sighting range, elevation	degrees	- 11 + 43
Sighting range, azimuth	degrees	1
Weight in firing position	kg	2200
Projectile weight	kg	42
Muzzle velocity	m/sec	365
Maximum range	km	8.5
Gun carriage		box-trail
Protective shield		yes
Locomotion		drawn by horse
In service in		Switzerland
Notes		Transported in one piece.

Photo (above) and specifications of the sFH 17.

sFH 17

The designers at Krupp and Rheinmetall began design work on the new Heavy Field Howitzer 18, for a new armed forces to be built up in the future, some months before the return of the International Control Commission on February 28, 1927. Three different sizes were envisaged initially. The old sFH 13 was incapable of meeting any of the target specifications. The design team was the same one that developed the other new models for the armed forces. The barrel came from Krupp, the carriage from Rheinmetall. With a barrel length of 4.397 meters, the new gun was more than twice as heavy as its predecessor, however muzzle velocity was increased fifty percent and firing range was upped to 13,300 meters. By existing standards the sFH 18 was a cannon-howitzer. But the new split-trail gun carriage, which finally made possible three-point seating and which approached the outrigger-type gun mount, increased the gun's range of traverse twelvefold!

Following Germany's departure from the League of Nations and the one-sided disarmament conference of October 14, 1933, the way was clear for a step-by-step reformation of the army's heavy artillery with modern weapons. The first artillery regiments were activated on October 1, 1934; as before, the fourth battalion of each regiment was set aside as the heavy battalion. The long period of secret testing allowed the manufacturers to immediately deliver a trouble free weapon to the field forces. The model number "18" was an attempt at camouflage. Because of its weight, more than 5.5 tonnes in firing position, when horse-drawn the sFH 18 was moved in two parts with separate wagons for the barrel and gun carriage, each part weighing about the same as the lFH 18 light howitzer. When it was towed by motor vehicle the weapon was not broken down. Allocating just the motorized version to the mass of the field army, as was the case with the regular army artillery battalions, was out of the question due to the raw materials situation. Its mobility, firing range and effectiveness with a projectile weight of barely 100 pounds made the sFH 18 the main weapon of the army division, of which it was, for the first time, an integral part.

sFH 18

The Reich War Ministry's official Wehrmacht postcard of 1935 unintentionally revealed the great difficulties of the two-part, horse-drawn gun – even with eight-horse teams, as in this case –in terrain other than firm roads. The troops were urged to overcome such problems through "a combined effort."

The horse-drawn sFH 18, which was introduced in preference to the lFH 18, in firing position on the practice range. Numerous breaches of standing orders are evident: K 1 wearing his steel helmet, K 2 with open collar, trail spades not dug in, gun carriage wheels not on the mats.

Stoffgliederung 5	Geheime Kommandosache! sFH 18 (Kzg)	Blatt G 122

Dringl.-St.: SS (sf s.: Kfz. Bl. G 367)

Technische Daten:

Schußweite 13325 m

Rohrweite 14,9 cm

Rohrlänge in Kalibern L/29,5 = 4,440 m

Geschoßgewicht 43,5 kg (15 cm Gr. 19)
25 " (15 cm Gr. 39 Hl/A)
15 " (15 cm Pzgr. 39 TS)

Größte V_0 520 m/sek (Gr. 19)
465 " (Gr. 39 Hl/A)
880 " (Pzgr. 39 TS)

Mündungswucht 600 mt

Zahl der Ladungen 8

Gewicht in Feuerstellung 5512 kg

Gewicht in Fahrstellung (1 Fahrlast) 6304 kg

Richtfeld Seite 60° Höhe 0° bis +45°

Schuß/min ~ 4

Geschätzte Lebensdauer eines Rohres 15-25000 Schuß (m. Ladungsraumfutter)

Durchschl.-Lstg.: auf 100 m = 155 mm; auf 1000 m = 126 mm (Pzgr. 39 TS)
(s. Vorbm. I) 160 mm starke Panzerplatte m. 15 cm Gr. 39 Hl/A

Geschütz-Verwertung (Mündungswucht mt / Geschütz-Gewicht t) ...... 108,5 mt/t ......

Rohstoffbedarf	Fe	Mn	Cr	V	Mg	Sn	Cu
f. Kfz. 1 Stck. i. kg	11482	85,2	0,065	4,4			0,32
m. Zubehör	Al	Pt	Zn	Ni	Kautschuk		
f. Kfz. 1 Stck. i. kg	0,003		0,83	27,0	138,-		

Gruppenpreis I RM 38500.- / " II " 40400.-	Durchschn. Fertigungszeit 9 Monate	Arbeitsstunden 5500

Fertigungsfirmen:	Geschoß:
Spreewerk, Bln.-Spandau	15 cm Gr. 19
MAN, Augsburg	15 cm Gr. 19 Stg.
Dörries-Füllner, Bad Warmbrunn	15 cm Gr. 36
Skoda, Dubnica / Slowakei	15 cm Gr. 19 Nb
	15 cm Gr. 38 Nb
	15 cm Gr. 40 Nb
	15 cm Gr. 19 Be
	15 cm R.Gr. 19
	15 cm Gr. 19 Br.
	15 cm Gr. 19 (Üb.B.)
	15 cm Gr. 39 Hl/A
	15 cm Gr. 19 Stg. (Üb.T.)
	15 cm Pzgr. 39 TS
	15 cm Spgr. 42 TS

The Reich Minister for Armaments and War Production included the sFH 18, although only the version with prime mover, in his 1944 "Specification Sheets for Army Weapons, Vehicles and Equipment." Based on experience, the deletion of the horse-drawn version was long overdue, even if it was only assigned the "SS" priority level. The split-trail gun carriage improved the range of traverse to 60 degrees. The unnecessary gun shield was dropped and maximum firing range was increased by 5,000 meters, or more than 60%. The multitude of propellant charges ensured a very versatile path of flight while the number of different projectiles allowed gunners to vary the effect of their fire.

An sFH 18 of III Battalion, 18th Motorized Artillery Regiment in Staraya Russa 1942-43. In spite of the bitter cold the battalion's two heavy field howitzer batteries provided reliable support in defending against continuous attacks from all sides. The winter camouflage was whitewash, which could be washed off at any time.

s. F. H. 18, Kzg.
Ansicht von vorn

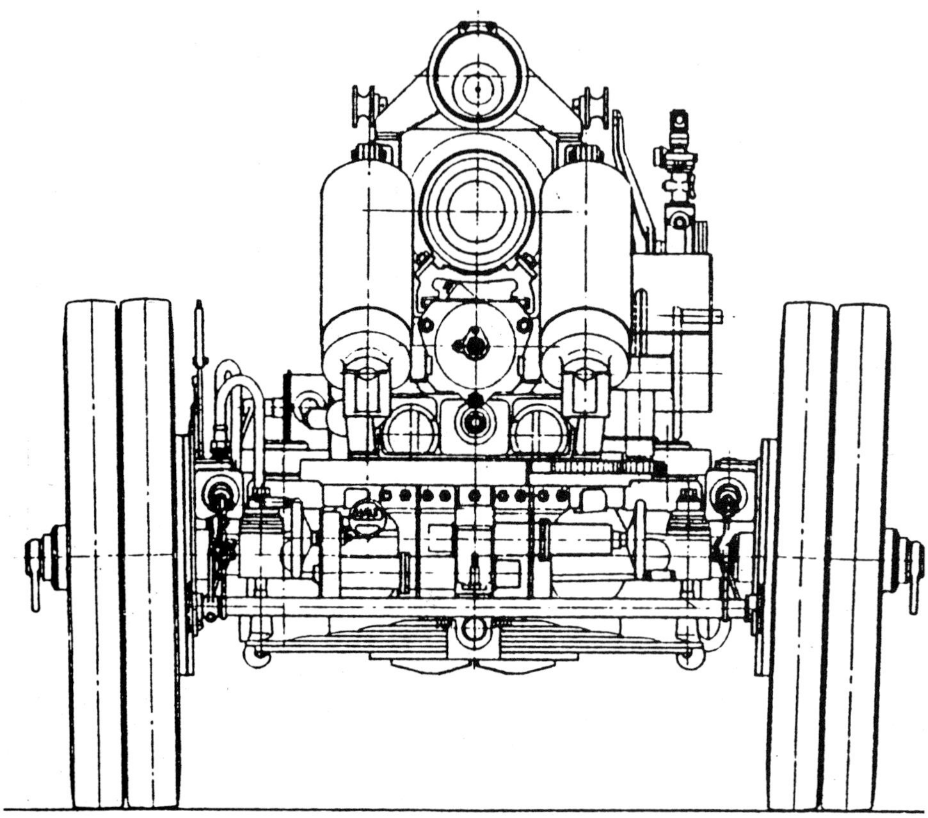

The following line-drawings of the sFH 18 are taken from the last example of the H.Dv.201/1 and 2 "Description and illustration of the sFH 18 for horse team and prime mover and the 100mm Kanone 18 for horse team and prime mover." Printed on May 1, 1940, it remained valid until 1945. It provides a good illustration of the most important structural components of the guns.

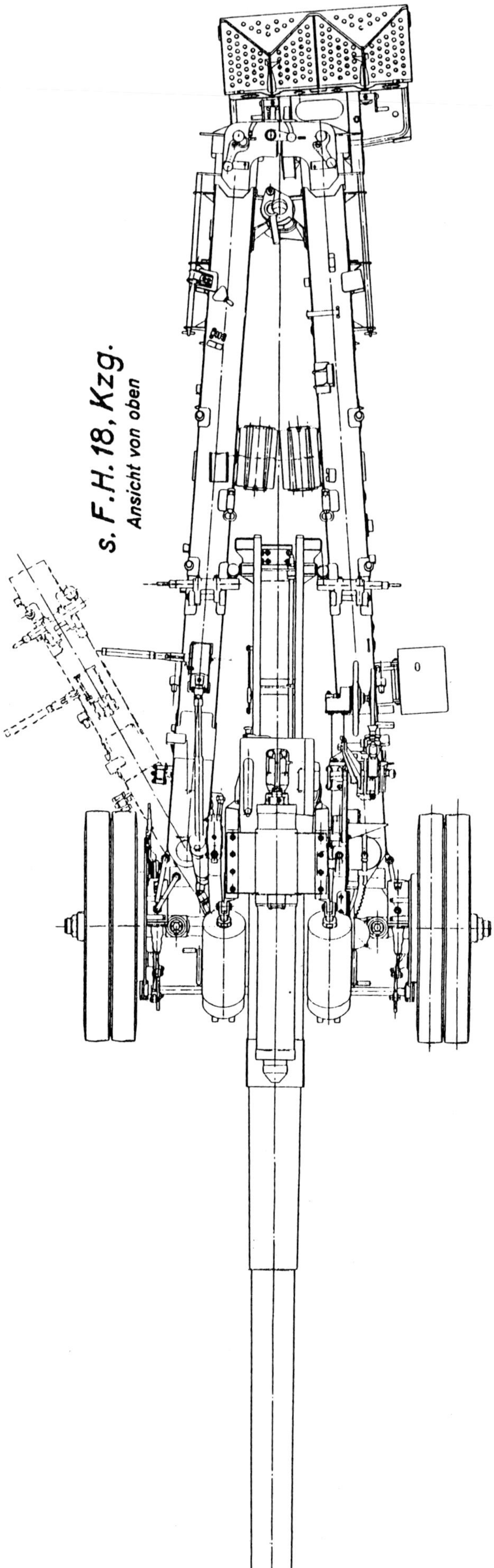

s. F. H. 18, Kzg.
Ansicht von oben

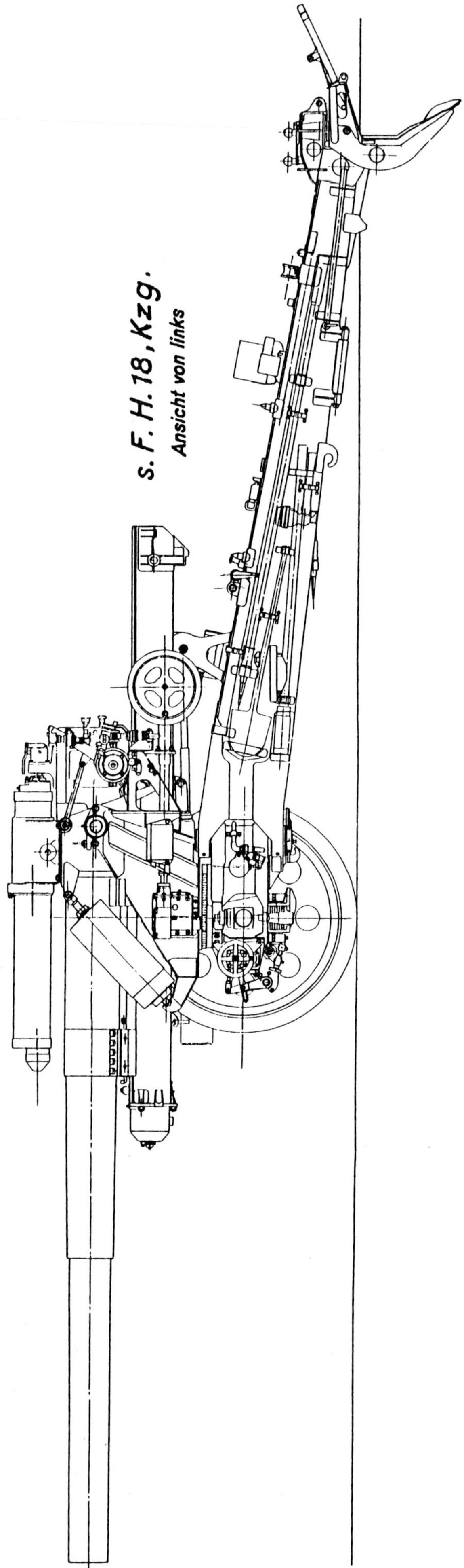

s. F. H. 18, Kzg.
Ansicht von links

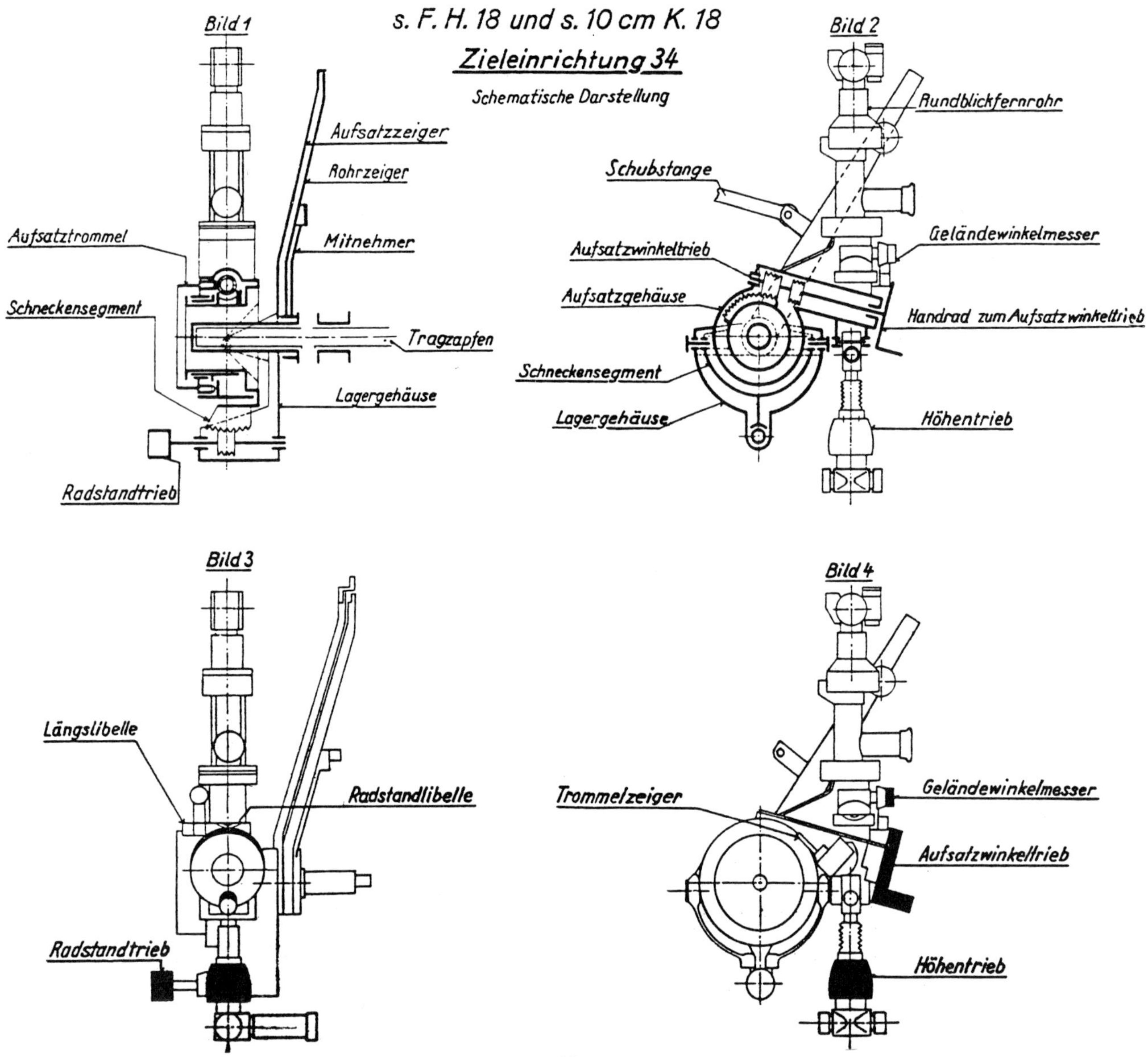

The introduction of the new Zeiger-Zieleinrichtung 34 (Pointer Sighting Mechanism) simplified orienting the gun and setting the elevation; the clear visual control system simplified cooperation with the 2nd Gunner (Kanonier 2), whose job it was to fire the gun.

Below:
In the beginning there was the horse-drawn sFH 18. It was lighter than the later variant designed to be towed by a prime mover and was generally transported in two pieces.

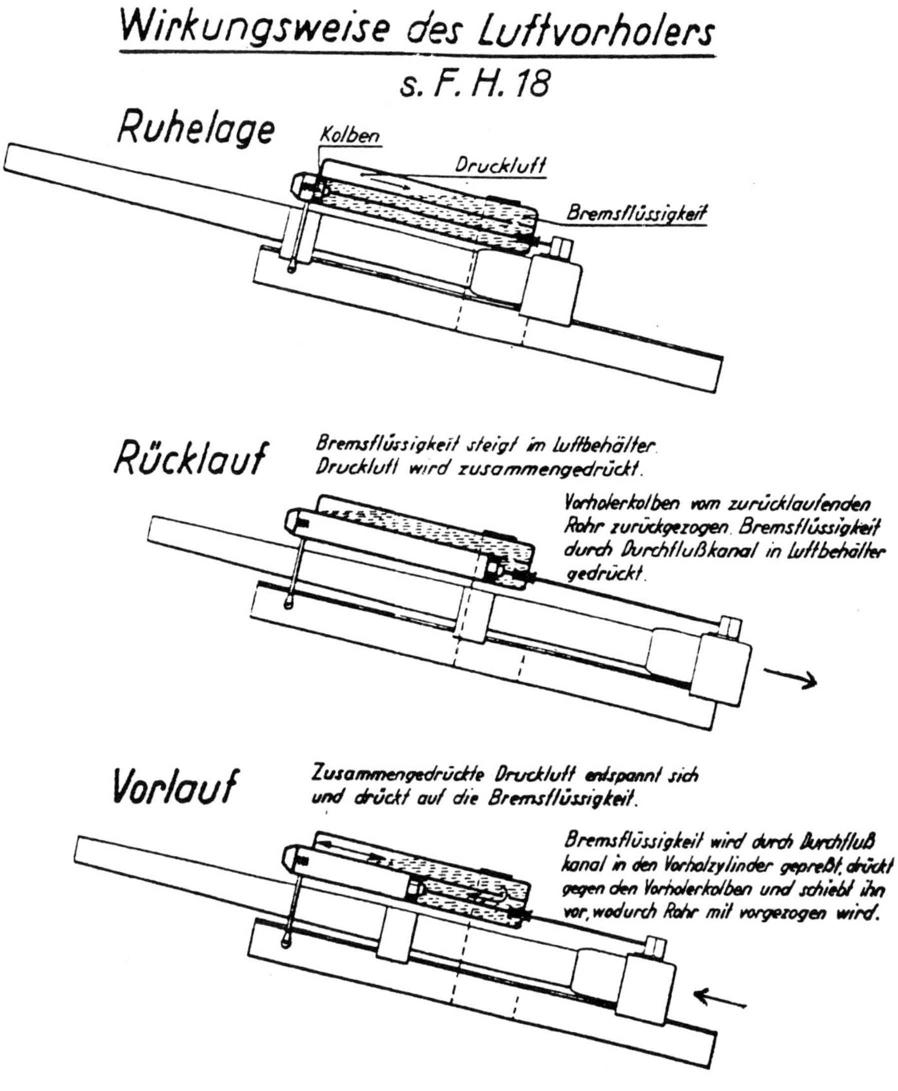

Above:
With the elevation set, the gun layer (K 1) checks the azimuth once again. To his right the K 2 (2nd Gunner) watches expectantly.

Left:
The working of the compressed air recuperator. Compressed air and brake fluid worked against each other alternately, in such a way that a doubled effect was achieved.

The horse-drawn version of the sFH 18, which served mainly in the 4th Battalion of artillery regiments of infantry divisions, was transported in two pieces. In this photo, taken in September 1939, four gun carriage trailers move into a firing position, in this case with the teams still hitched, in preparation for opening fire. Transporting the weapon in two parts did little to improve its mobility.

Horse-drawn sFH 18

The barrel trailer on the right has been moved close to the gun carriage trailer on the left in order to immediately attach the barrel. Winter 1939-40 at the Westwall.

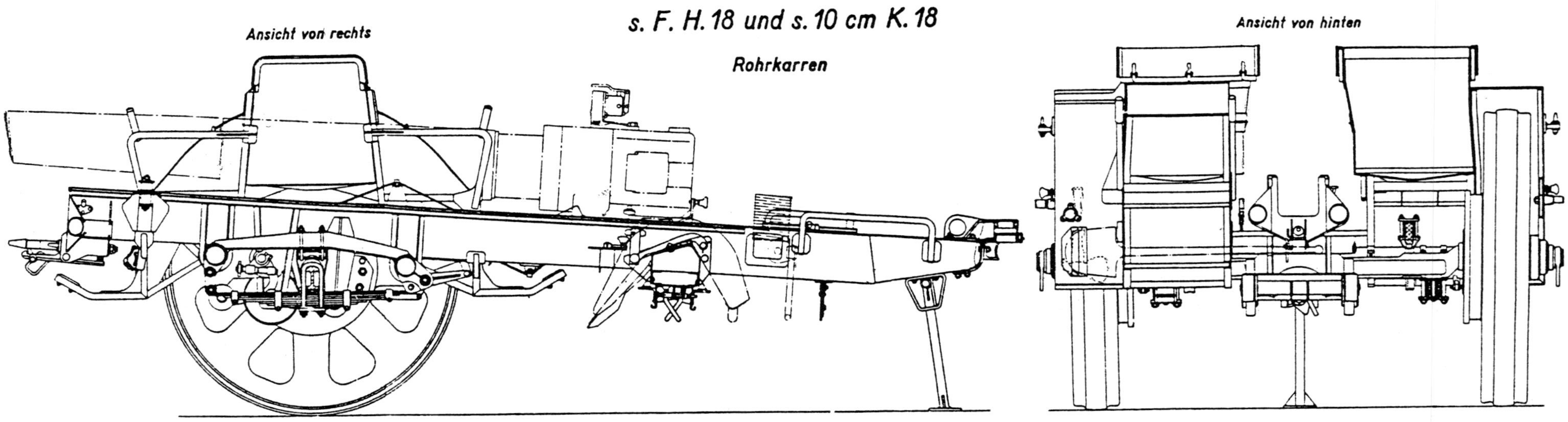

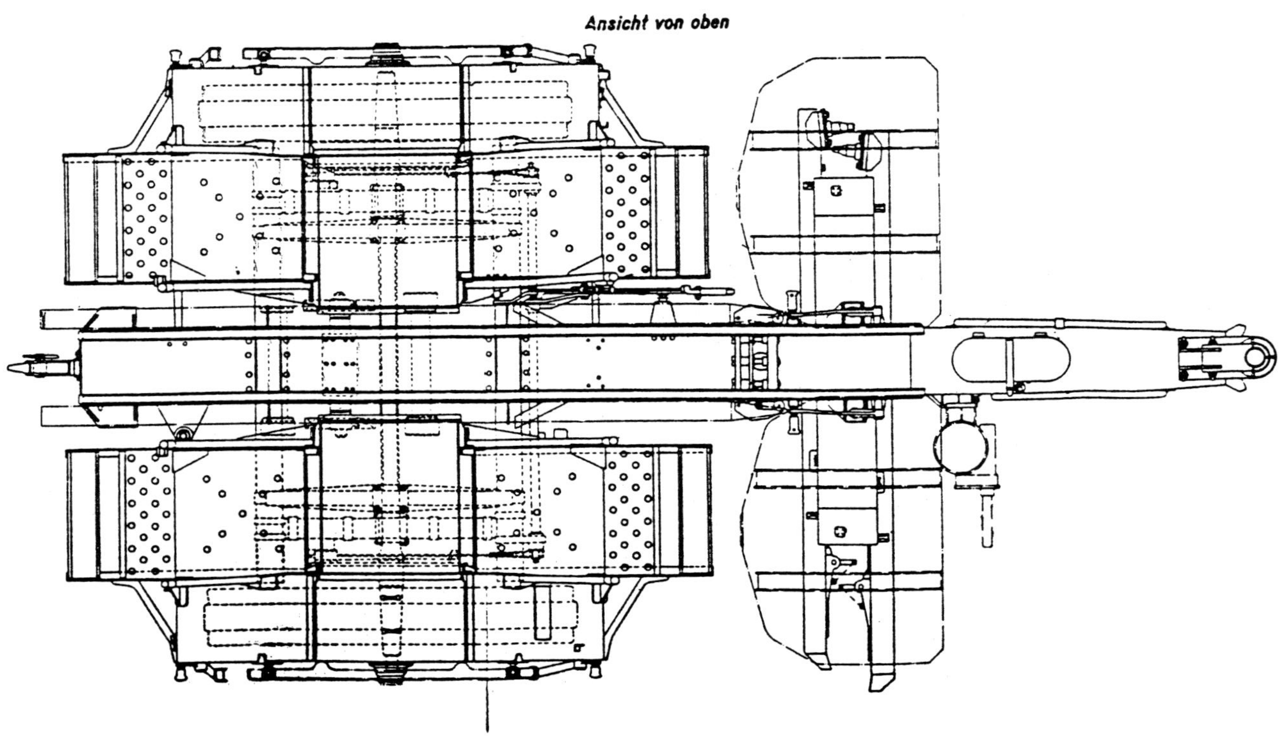

The barrel trailer, also called the "barrel cart," transported just the 4.44-meter-long barrel. The barrel was positioned low, almost right over the trailer axle, and was drawn back only far enough to maintain the trailer's center of gravity. The technique of moving the weapon in two parts naturally made it more difficult and time consuming to achieve firing readiness; as well it added to the gun crew's workload.

Above: The gun trailer is manhandled up to the gun, which is in firing position, in preparation for movement to a new position. Once this is done, the gun carriage will be fetched from the team.

Below: Loading ramps were used to bridge the difference in height between the gun trailer and the firing height of the barrel on the gun carriage, which was significantly greater. In preparation for removing the barrel as part of a change of position, the barrel trailer was positioned to the right of the gun carriage trailer.

s. F. H. 18 und s. 10 cm K. 18

Rohrkarren, Auffahrschienen

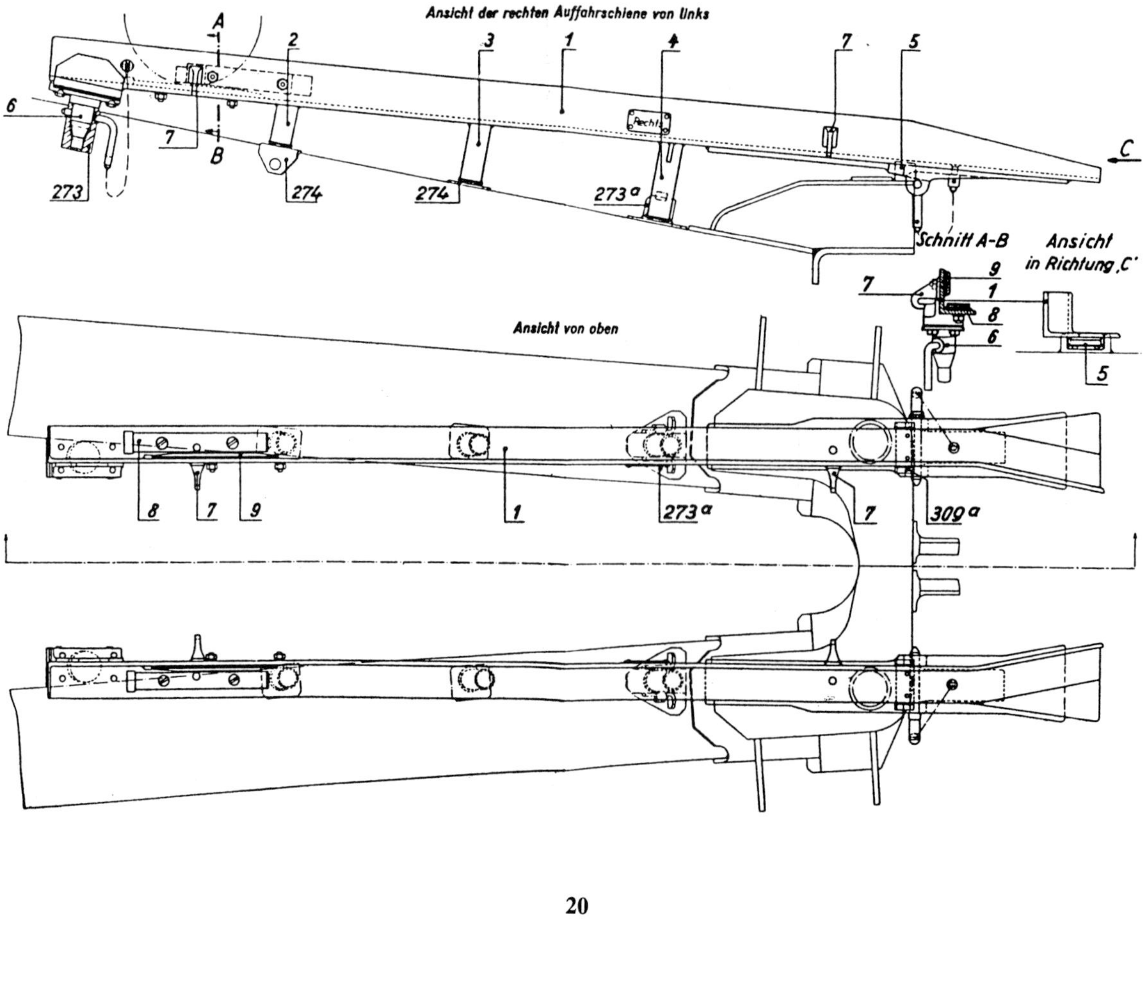

New Models Developed During the War

It was years, until 1942, before the war situation finally forced the designers to produce improvements or new designs. The objective of these was primarily to lighten the weapon, but also to increase its range. The sFH 36 L/23 was developed in parallel by Krupp and Rheinmetall from 1936 to 1939. The gun's barrel was 99 centimeters shorter than the sFH 18 and its range was decreased by 825 meters to 12,500, also there was a muzzle brake for the first time. They managed to reduce transport weight by 2.8 tonnes to 55% of the sFH 18, and in firing position by 2.23 tonnes to 59.6% of its predecessor, so that the horse-drawn version could be transported as a single load with the help of a bogey trailer. Work on the project was halted in 1942, because the sFH 40, the specification for which had been issued in 1938, promised better results. True, it was 100 kilograms heavier than the sFH 18 in transport and firing positions, but with its muzzle brake and a barrel lengthened by 450mm, using a No. 9 charge it could lob a shell 15,600 meters, or more than 3,175 meters more than the sFH 18. As well, it was capable of high-angle fire, with elevations of up to 70 degrees. Further improvements were offered in the form of a pneumatic brake instead of the spring barrel brake, variable barrel recoil and an enlarged powder chamber in the barrel. All in all, it was an new and improved version of the sFH 18 without the costs of a new development. However, production was not started in 1942, because the OKH was demanding increased production of the Heavy 100mm Kanone 18 and the sFH 18, and this would suffer through a changeover to the new model. As well there were raw materials problems. The designers subsequently combined the barrel of the sFH 40 with the gun carriage of the sFH 18. By using an extremely powerful muzzle brake and a No. 9 charge it was possible to achieve a range of 15,100 meters, while weight in driving and firing positions was equivalent to that of the sFH 40. The result was the sFH 18/40. Only 46 examples of this acceptable compromise, which was developed in 1942-43, were built. One new feature was the self-obturating, sliding-wedge-type breechblock, which did not require steel cartridges. This variant was later also called the sFH 42 and 18/43. Krupp, Rheinmetall and Skoda all came up with breakthroughs in gun carriage design in 1942-43, pointing the way toward the modern three-spar and outrigger-type carriages for all-round fire. However, these proceeded no farther than wooden models, because from 1942 on all development work in the field of artillery was suspended in favor of increased production of armored vehicles of all kinds.

These gun carriages, for which there were several designs, made it possible to fire in the upper angle ranges to a maximum of 70 degrees, a range of 18 kilometers with a weight reduction of 500 kilograms in firing position and a weight increase of the same amount in the transport position as well as the simultaneous use of the lower carriage for the 128mm Kanone 43. After the war all the victorious powers, especially the French, continued development of these German designs based on Eastern Front experience. Although development of a modern, medium field artillery piece was entirely feasible, pressures caused by the war situation prevented it. Times have changed. Today in the Bundeswehr a weapon with a calibre of 155mm is considered "light field artillery" and it has tanks with 105mm and 120mm guns.

Winter 1943 at the Mius River, shortly before the German Army's withdrawal.

sFH 18 mit Zgkw.

Stoffgliederung 21	Geheime Kommandosache! m Zgkw 8t (Sd Kfz 7)	Blatt G 402 V

Dringl.-St.: SS

Technische Daten:

Gesamtgewicht des Fahrzeuges (Gefechtsgewicht) 11,55 t
Anhängelast normale = bis 8 t, Gesamtlastzug = 19,55 t
Motor: Maybach HL 62 TUK 140 PS (Fahrg.-Gewicht = 7,95 t
Verhältnis PS zu t = 12,12 PS/t
Länge 6,85 m
Breite 2,40 m
Höhe m. Verdeck 2,62 m
Sitze vorn 3 hinten 8
Höchstgeschwindigkeit 50 km
Steigfähigkeit in losem Sand mit Last 12°
" " " " ohne " 24°
Bodenfreiheit unter d. Vorderachse 0,415 m
Watfähigkeit 0,65 m
Grabenüberschreitfähigkeit 1,80 m
Mitgeführte Kraftstoffmenge 200 l
Fahrbereich mit einer Kraftstoffüllung (einschl. Reservetank)
Straße: ~ 250 km; mittl. Gelände: ~ 135 km

Kette: 54 Glieder Ketten-Typ: Zgw. 50/360/180 = 635 kg
Anhängelast f. Sd. Kfz. 7: a) s. F. H. 18
b) s. 10 cm K.
c) 8,8 cm Flak

Fahrgestellverwendung		dazugehörige Anhängelast
Sd. Kfz. 7/2	3,7 cm Flak 36 Sfl.	Sd. Anh. 57
7/1	Sfl. f. 2cm Flak Vierling 38	Sd. Anh. 56

Kette	Glieder, Kettengewicht	kg	Ni	Kautschuk (Reifen + Polster)	
				291,6	

Rohstoffbedarf f. 1 Stck. i. kg	Fe	Mo	Cr	W	Mg	Sn	Cu	Al	Pb	Zn
	16000					2	50	41	68	35

Preis RM 36000.–	Durchschn. Fertigungszeit 15 Monate	Arbeitsstunden 2050 f. mech. Bearbtg. 1450 f. Montage

Fertigungsfirmen: Krauss-Maffei A.G., München-Allach
Carl F. W. Borgward, Bremen
Österr. Saurerwerke A.G., Wien

Above: Limbering and unlimbering were considerably easier when the gun was transported by prime mover: the closed split-trail gun carriage sat on a bogey trailer, insertion of which placed the heavy gun in the proper balanced position. The barrel was pulled back into the transport position to improve the center of gravity situation and reduce the gun's overall length during transport. The Krauss Maffei KM m 11, an 8-ton, half-tracked prime mover, was a very reliable gun tractor.

Near Kursk in 1943, a gun crew struggles to turn its sFH 18 in the proper firing direction. The man hanging on to the muzzle is acting as a sort of counterweight. The mats under the wheels are supposed to prevent them from sinking into the earth. With a weight of more than 6 tons, the crew had to use balance and leverage in order to turn the gun by hand.

Below:
Photo taken during trials with sleigh-runners attached to the gun carriage and bogey trailer at St. Johann, Tyrol on January 13, 1942.

sFH 18 with Heavy Wehrmacht Tractor (SWS)

Later in the war the 8-ton Heavy Wehrmacht Tractor also served as a means of transporting the sFH 18, as here near Kursk in the summer of 1943. When moving the weapon into position, the crew moved the barrel forward into firing position before unlimbering. The leverage provided by the extended barrel made it easier to unlimber the weapon.

Below: In spite of extremely difficult ground conditions, the gun crew couple the gun carriage to the Heavy Wehrmacht Tractor again before moving to a new position; the barrel is already in the transport position.

Stoff-gliederung 21	Geheime Kommandosache! sWehrmachtschlepper (Typ. s WS)	Blatt G 402 IV

Technische Daten:

Dringl. - St.: DE

Gesamtgewicht d. Fahrzeuges (Gefechtsgewicht) = 13,5 t
Anhängelast normale = 8 t
Gesamtlastzug = 21,5 t
(Fahrgest. - Gewicht = 8 t

Motor = 100 PS
Verhältnis PS zu t 7,4 PS/t

Länge 6,68 m
Breite 2,50 m
Höhe m Verdeck 2,83 m

Sitze: 2 + 10

Höchstgeschwindigkeit 27 km
Steigfähigkeit i. losem Sand m. Last 12°
" " " " ohne " 24°

Bodenfreiheit 0,50 m
Watfähigkeit 1,00 m
Grabenüberschreitfähigkeit m

Mitgeführte Kraftstoffmenge 240 l

Fahrbereich m. einer Kraftstoffüllung (einschl. Reservetank)
Straße 300 km; mittl. Gelände 100 km
Kette: 110 Glieder ; Ketten - Typ. Gg 25500/180 = 1050 kg (eine Kette)

Fahrgestellverwendung: Sfl 3,7 cm Flak

Official data sheet for the Heavy Wehrmacht Tractor. In this case the vehicle was used as a self-propelled carriage for a 37mm Flak.

Kette Glieder, Kettengewicht kg							Ni	Kautschuk		
Rohstoffbedarf f. 1 Stck. i. kg	Fe	Mo	Cr	W	Mg	Sn	Cu	Al	Pb	Zn

Preis RM	Durchschn. Fertigungszeit Monate	Arbeitsstunden

Fertigungsfirmen: Büssing NAG, Berlin - Oberschöneweide
Ringhofer - Tatra Wagenwerk Kolin

Above and below:
On the retreat from Kursk at the beginning of July 1943 and in Northern Hungary in August-September 1944. Introduced in 1943, the Heavy Wehrmacht Tractor was 2 tons heavier than the 8-ton prime mover and its tracks were twice as heavy. With a less-powerful engine, the top speed of the SWS was only half that of the earlier prime mover, but it possessed a better fording ability and greater range on roads.

Stoff-gliederung 5	**Geheime Kommandosache!** sFH 42	Blatt G 123

Dringl.-St.:

Technische Daten:

Schußweite 15 100 m

Rohrweite 14,9 cm

Rohrlänge in Kalibern L/32,5 = 4,875 m (o. Mdgsbr.)
L/36 = 5,388 m (m. ")

Geschoßgewicht 43,5 kg

Größte V_0 595 m/sek

Mündungswucht 784,9 mt

Zahl der Ladungen 9

Gewicht in Feuerstellung 5660 kg (m. Lm.-Rädern = 5470 kg.)

Gewicht in Fahrstellung (1 Fahrlast) 6480 kg

Richtfeld Seite 56° Höhe 0° bis +45°

Schuß/min ~4

Geschätzte Lebensdauer eines Rohres ~10 000 Schuß

Durchschl.-Lstg.:

Lt. Befehl d. Führers ist die Schußweite der sFH 18 bis auf 16 km zu steigern. Fa. Krupp hat die Aufgabe gelöst, Versuchsgerät u. Lehrauftrag ausgeliefert. Weiterfertigg. zurückgest.

Geschütz-Verwertung $\left(\frac{\text{Mündungswucht mt}}{\text{Geschütz-Gewicht t}}\right)$ 138,6 $\frac{\text{mt}}{\text{t}}$

Rohstoffbedarf	Fe	Mo	Cr	W	Mg	Sn	Cu
f. 1 Stck. i. kg	19 120		54,8				8,-
	Al	**Pb**	**Zn**	**Ni**	**Kautschuk**		
f. 1 Stck. i. kg	30		4,8	42	138,-		

Richt Preis RM 60.000,-	Durchschn. Fertigungszeit ~12 Monate	Arbeitsstunden ~9000 (geschätzt)

Fertigungsfirmen:	Geschoß:
Spreewerk, Bln.-Spandau MAN, Augsburg Dörries-Füllner, Bad Warmbrunn Skoda, Dubnica / Slowakei	15 cm Gr. 19 15 cm Gr. 19 Stg. 15 cm Gr. 36 15 cm Gr. 19 Nb 15 cm Gr. 38 Nb 15 cm Gr. 40 Nb 15 cm Gr. 19 Be 15 cm R.Gr. 19 15 cm Gr. 19 Br. 15 cm Gr. 19 (Üb. B.) 15 cm Gr. 39 Hl/A 15 cm Gr. 19 Stg. (Üb. T.) 15 cm Pzgr. 39 TS 15 cm Sprgr. 42 TS

Type sheet for the sFH 42, which entered service in 1943, issued on January 1, 1944. The handwritten note halfway down the page states: "On the Führer's order the firing range of the sFH 42 is to be increased to 16 km. The Krupp Firm has delivered prototypes and training material. Further production deferred."

Above: Moving to the next firing position. Because of the mountainous terrain on the border with Southern Russia, the barrel trailer is, for a change, being pulled by an eight-horse team.

Into firing position

Below: The gun carriage trailer drives into position while the rest of the battery follows. The relative lack of experience of the drivers is evident in their arm movements.

There were also often difficulties when – as here in Northern Russia in August 1944 – a battery of heavy field howitzers had to be moved across a lake, gun by gun, using a ferry consisting of a platform over four pontoons.

Moving into a firing position off the village street in Przystawy in Poland in September 1939. Two years later no one would have bothered to smooth the road.

Above:
The gun has reached the firing position. Having unlimbered the gun, the gunners are removing all necessary items of equipment from the gun tractor before it drives off into the limbering position. The driver waits for his comrades to complete their task.

Left:
The instrument NCO in charge of the aiming circle then surveys the gun in order to ensure that it is parallel to the base point. Each gunner represents an azimuth reading.

The battery officer was the soul of the firing position, which he supervised and commanded. He was responsible for the proper carrying-out of firing commands and fire orders. Most were young officers, but their ranks also included experienced senior technical sergeants (Oberwachtmeister) like the man in this photo. Chest microphone and firing tables were his most important tools.

Left:
For each two guns the battery officer was supported by the platoon leader, Wachtmeister (Technical Sergeant) or more senior NCO, who repeated the commands out loud and worked with a chest microphone and firing tables. This extravagant command apparatus with the gun commanders was necessary to ensure that the battery commander's commands were strictly and properly carried out and that fire discipline was maintained.

Right:
The regiment's meteorology platoon provided the necessary data concerning temperature, wind, barometric pressure and so on. The heavy artillery in particular had to work closely with it. Here a gunner takes the surface temperature. Weather balloons were used to measure temperatures aloft.

The first three guns of a heavy field howitzer battery stand ready to fire on a firing range at dawn. In command of the battery is a Technical Sergeant, standing on the right wearing the peaked cap.

Another heavy field howitzer in firing position on the range. The gun commander directs the 1st and 2nd Gunners, who are wearing steel helmets. Only a gesture has been made toward camouflage, the men are still wearing crests on their helmets and the position is neat and tidy.

Below: A combat firing position looks completely different. Such a scene would have provoked disciplinary measures in peacetime. This photo was taken during a pause in firing after a barrage. Heat and hard physical work have led the crew to remove their shirts.

Preparing a heavy field howitzer to fire, winter 1942. The cover is removed from the sighting mechanism, the breach is opened and the snow is removed from all working parts.

The photo above shows an sFH 18 in an emplacement prior to the Battle of Kursk in 1943. Everything has been unpacked and laid out. On the left is the bogey trailer, while the battery officer's car has been driven into cover off to the side. A guide is apparently giving the crew more detailed instructions.
The lower photo shows the loading process prior to inserting the shell. In this pre-war exercise the gun crew are wearing gas masks, something that scarcely ever done in action.

The 43.52-kg projectile, almost a hundredweight, had to be pushed forcefully into the barrel with the long rammer. Then followed the cartridge case with the specified number of charges; the man on the left has it ready. The steel helmet on the head of the gun layer suggests that this is a peacetime exercise.

Left:
While the gun layer once again checks the base-line direction, the projectile is rammed in, the cartridge follows immediately and the next shell stands ready – a scene from a live-firing exercise at Neufchatel, France on April 4, 1941 prior to the invasion of Russia.

Below:
A gunner about to fire an sFH 18 in the dawn of the first Russian winter. In a second there will be a loud crash as the gun is fired. The gun crew are wearing loose-fitting driver's coats, which were supposed to take the place of the missing winter clothing.

The gun crew has stepped back, the gun commander is standing to the right. The K 2 fires the sFH 18 using the pull-cord, probably live firing as part of an exercise.

Above: September 1939. During a break in the firing a gunner prepares the shells for the next salvo. Still inexperienced, the crew has laid the projectiles on the ground, which was only possible for a brief time and in dry weather.

Below: 32 rounds, which the platoon commander has just had delivered by the ammunition section, are laid out for a counterattack or a concentration of fire at dawn on February 6, 1944. While the platoon commander jots down his mission, the gunners are already setting the fuses.

Ammunition was also laid out to the side of the firing position. Here two Obergefreiters (Lance Corporals) prepare them for use. These rounds are grey-green-painted FES 36 high-explosive shells, which were somewhat longer than the other ammunition. September-October 1943.

New Developments

This significantly lighter heavy field howitzer was designed to be moved in one piece by horse-team, but at the cost of a reduction in range of 1,000 meters. Under development since 1936-1939, it was discarded in favor of the sFH 40 in 1942.

The sFH 40, under development since 1938, incorporated some real advances in design: an increase in range of more than 2,000 meters and 70-degree elevation. Prototypes were produced in 1941, but development was halted in 1942 in order to produce more sFH 18 and 100mm Kan 18 weapons.

Below: The sFH 18/40 was a compromise, utilizing the barrel of the sFH 40 and the carriage of the sFH 18. With a weight of 6.5 tons, the gun possessed a firing range of more than 15 kilometers. Developed in 1942-43, only about 46 of these guns were built.

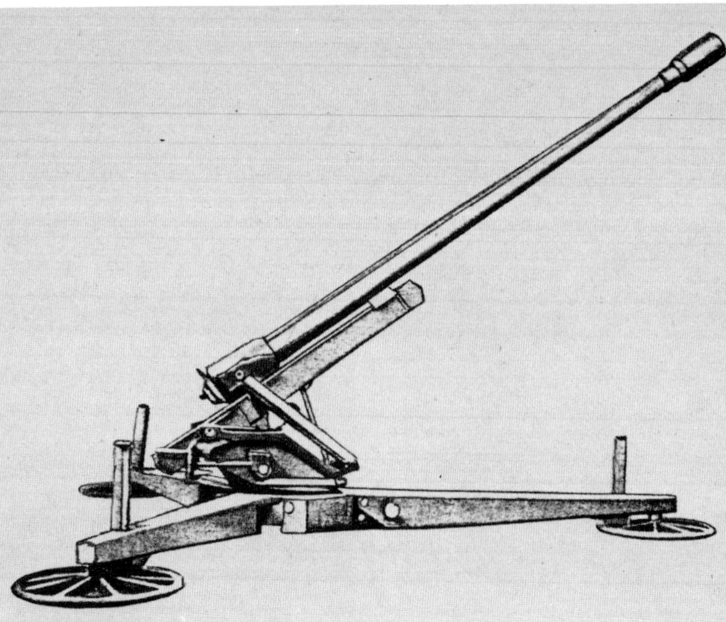

15 cm s. FH 43. Erster Entwurf Krupp mit Dreiholmlafette.

15 cm s. FH 43. Zweiter Entwurf Krupp mit Kreuzlafette in Fahrstellung

Left:
Like the leFH 43, versions of the sFH 43 were proposed with three-spar and outrigger-type gun carriages for all-round firing and firing at extreme angles of up to 70 degrees. This was all part of an attempt to develop a standard gun carriage. Three firms participated in the effort.

Right:
The designs were finalized in 1942-43. Wooden mock-ups were built at the end of 1942; however the project went no further since at the end of 1942 development of guns for the artillery was dropped in favor of increased tank production. The sFH 43's design range of 18 kilometers exceeded that of the Russian 152mm cannon-howitzer by 1,000 meters. After the war France continued development of this new concept for a modern medium artillery piece.

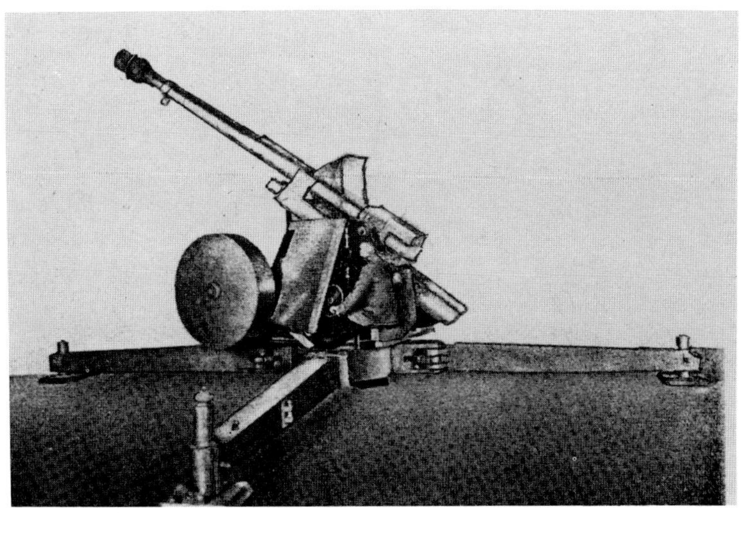

Hummel (Bumblebee)

The third possible form of locomotion for the heavy field howitzer, a self-propelled carriage, was realized in the GW III/IV "Hummel" with the sFH 18/1. While externally it had the visual appearance of an armored howitzer, it was in fact merely a self-propelled gun whose fighting compartment had lightly-armored sides, offering the crew protection against small-arms fire, but with an open top.

Stoffgliederung 21	**Geheime Kommandosache!** s Pz Haub 18/1 auf Fgst. Pz KpfW III/IV (Sf) (Sd. Kfz. 165)	Blatt G 367

Dringl.-St.: SS

Technische Daten:

Gesamtgewicht des Fahrzeuges (Gefechtsgewicht) 24 t

Motor HL 120 265 PS bei 2600 U/min

Höchstgeschwindigkeit 42 km/Std.

Mitgeführte Kraftstoffmenge 470 l (einschl. Reservetank)

Fahrbereich mit einer Kraftstoff-Füllung:

Straße 215 km; mittl. Gelände 130 km

Grabenüberschreitfähigkeit 2,3 m

Besatzung Fahrer + 1 + 4

Länge 7,170 m, Breite 2,970 m

Höhe mit Aufbau 2,810 m — 3,260 m m. Ostkette

Bordmunition 18 Schuß f.s.F.H. 18

Bestückung: a) Turmwaffen 1 s.F.H. 18

b) Bugwaffen

Abfeuerung

Optisches Gerät: a) Turmoptik 1 Rbl. F. 36

b) Kugeloptik

c) Fahreroptik

Funkgerät (normale Ausstattung) Fu Spr „F" u. Bord Spr.-Gerät

Panzerung: Front 20 mm Seite 20 mm

Turm 10 mm Dach

Kette 104 Glieder, Kettengewicht 805 kg; Ostkette 1510 kg

	Ni	Kautschuk (Reifen usw.)	
		125	

Rohstoffbedarf f. 1 Stck. i. kg	Fe	Mo	Cr	W	Mg	Sn	Cu	Al	Pb	Zn
	37000				–	1,5	193,1	295	62,4	66

Preis RM	Durchschn. Fertigungszeit 8 Monate	Arbeitsstunden

Fertigungsfirmen:

Montage u. Fahrgestelle: Deutsche Eisenwerke, Duisburg

Panzerung: Deutsche Röhrenwerke, Mülheim

Technical Data

Modell	lg. s. FH 13	15 cm s. FH 18	15 cm s. FH 36	15 cm s.FH 18/40(42)	15 cm s. FH 40	s. FH 43
Kaliber mm	149,7	149	149,1	149	149,1	149
Rohrlänge mm	2550 L/17	4440 L/29,5	3555 L/23.7	4875 L/32,5	4875 L/32,5	4730 L/31,5
Seelenlänge	–	3985	2965	-	–	–
Gezogener Teil mm	2054	3623	2475	3927	3927	–
Verbrennungsraum	173	362	341	506	506	–
Verbrennungsraum Liter	3,2	7,22	6,8	9,85	10,1	–
Züge Zahl	–	40	40	40	40	–
Züge Tief mm	–	–	2,75	1,15	1,15	–
Züge Breite mm	–	–	7,00	6,71	6,71	–
Felderbreite mm	–	–	5,25	5,00	5,00	–
Drall, Art und Winkel	Steigender Rechtsdrall 4° zu 8°	Steigender Rechtsdrall 5° zu 10°	Steigender Rechtsdrall 6° zu 10°	Steigender Rechtsdrall 6° zu 9°	Steigender Rechtsdrall 4° 29' 27" zu 7° 9' 45"	– –
Abfeuerung	mechanisch	mechanisch	mechanisch	mechanisch	mechanisch	mechanisch
Rücklauf maximal mm	1500	1140	1160	920	1125	–
Rücklauf maximal mm	900	–	980	880	775	–
Rücklauf normal mm	–	1120	–	–	–	–
Richtmittel	– – –	ZE 34 mit Rbl.F. 16, 32, 36 oder 37	ZE 34 mit Rbl.F. 32, 36, 37 oder 38	ZE 34 mit Rbl.F. 32, 36 oder 37	– – –	– – –
Schwenkbereich°	25	30	28	28	30	360
Erhöhung°	0 bis + 45	0 bis + 45	- 1 bis +43	+ 45	- 1 bis + 70	- 8 bis + 70
Feuergeschwindigkeit S/min	4	4	4	4	4	4
Fahrstellung Länge mm	–	7849	10425	–	–	–
Fahrstellung Breite mm	–	2255	1752	–	–	–
Fahrstellung Höhe mm	–	1707	1880	–	–	–
Feuerstellung Länge mm	–	–	6300	5600	–	–
Feuerstellung Breite mm	–	–	4775	–	–	–
Feuerstellung Höhe mm	–	–	1780	–	–	–
Gewicht Fahrstellung kg	3000	6304	3500	6480	6400	7000
Gewicht Feuerstellung kg	2250	5512	3280	5660	5680	5000
V_0 maximal m/s	381	520	485	595	595	–
Gebrauchsgas-druck atü	2000 – 2550	2500	2600	2800	2600	–
Mündungswucht m/t	300	599,5	700	785	840	–
Schußweite maximal m	8675	13325	12500	15100	15675	18000
Zahl der Länge	8	8	7	9	9	8

MUNITION 15 CM SFH

Geschoßart	Brisanz	Beton	Hohlladung	Brisanz	Nebel	Rückstoß
	15 cm Gr. 19	15 cm Gr. 19 Beton	15 cm Gr. 39 FES-III/A	15 cm Gr. 36 FES	15 cm Gr 19 Nb	15 cm RGr 19
Gewicht kg	43,62	37,88	24,83	länger als 15 cm Gr. 19	38,97	45,25
Farbe	graugrün	durchweg dunkelgrün	feldgrau	graugrün	graugrün	–
Führungsring	2 Metall	2 Metall	1 Weicheisen	1 Weicheisen	2 Metall	–
Beschriftung	–	–	HI/A in Schwarz, FES in Weiß		Nb in Weiß auf	–
Zünder	AZ 23 umg 0,15 DoppZ S/60	BdZ –	klAZ 40 4 NbP –	AZ 23 0,15 DoppZ S/60	AZ 23 NB	El, AZ El, AZ mR DoopZ S/90
Verwendet bei Geschütz	15 cm sFH 18 nur I.-VII.Ldg. sFH 18/40(42) sFH 36	15 cm sFH 18 sFH 18/40(42) sFH 36 –	15 cm sFH 18 sFH 18/40(42) sFH 36 –	15 cm sFH 18 sFH 18/40(42)	15 cm sFH 18 sFH 18/40(42)	nur bei 15 cm sFH 18/M sFH 18/40(42)

Organizational and Technical Considerations

After the original rented civilian teams and the regular troops' own teams of horses, which made the artillery mobile, in 1928 the first motorized (tractor-drawn) heavy field battery was formed as a trials and evaluation unit. This led to the development of the various classes of Wehrmacht prime movers, which later proved so effective, with their classic half-track chassis consisting of a steering axle in front and caterpillar track drive in the rear. In 1931 an "Instruction and Test Detachment" was formed from two motorized heavy field howitzer batteries, motor transport battalions and training units. By 1935 there was already a total of 42 heavy battalions, and consequently the heavy artillery could be separated from the field artillery again. By 1937 there were 63 heavy battalions; with the divisions equipped, it was possible to form corps artillery units, to be used as reserve and main force artillery, the exact opposite route to that taken before the First World War.

By 1938 there were 70 heavy battalions and by the time World War Two broke out there were 88. This number climbed to 113 battalions in 1941 and 341 by the end of the war. During peacetime the annual formation rate was between 10 and 20 battalions. The total number of battalions was tripled in the last four years of the war, with captured weapons from all of Europe and coastal defence battalions adding significantly to the total. In 1942 the artillery regiments of panzer divisions were given a major boost in mobility with the introduction of the so-called "Hummel Battalion," a combination thought to be impossible only two decades earlier. By war's end there were 38 Hummel batteries. In spite of its success in action–after some initial problems–production was halted in 1944.

In all there were 14 different projectiles for the heavy field howitzer in all its variants. The sFH 18 cost between 38,000 and 40,000 Reichsmarks, the sFH 42 60,000. The average time required to build an sFH 18 was nine months, the sFH 42 twelve. Manufacturer was the Spree Factory in Berlin-Spandau, MAN in Augsburg, the Dörries-Füllner Werke in Bad Warmbrunn and Skoda branch factory at Dubnica in Slovakia. The life of an sFH 18's barrel was twice that of the sFH 42's, although the latter required almost twice the materials to build and as well had numerous accessories made from extremely scarce non-ferrous metals.

After evolving from stationary fortress artillery and ponderous foot artillery to modern heavy artillery in the field in a period of fifty years, the shortage of raw materials experienced during the war brought a certain end to the attempts by the manufacturers to produce an improved heavy field gun.

The Spielberger German Armor & Military Vehicles Series

Size: 8 1/2" x 11"   288 pages   hard cover
over 460 photographs
ISBN: 0-88740-397-2 $39.95

Size: 8 1/2" x 11"   256 pages   hard cover
over 240 photographs
ISBN: 0-88740-398-0 $39.95

Size: 8 1/2" x 11"   168 pages   hard cover
over 200 photographs
ISBN: 0-88740-448-0 $29.95

Size: 8 1/2" x 11"   168 pages   hard cover
over 200 photographs
ISBN: 0-88740-515-0 $29.95